# DIVERSITY ANALYSIS, HOLISTIC ENERGETICS, AND STATISTICS

The resonator complex model of the vegetation stand

All in one direction. Blind choices will not do …
On fieldtrip in Rio Grande do Sul with Professor Aino Jacques

**Front cover:** Campos landscape from the Rio Grande do Sul high lands

**Headliner:** Conceptual tools are presented in context, regarding their innovative use in community ecology.

**What is inside?** The book presents results from a conceptual analysis of disorder-based entropy (DBE) and energy-based entropy (EBE). DBE is the central disorder scalar in diversity analysis and EBE in holistic energetics. As such DBE and EBE are well conditioned, dialect-defining quantities in statistics. DBE and EBE are not interchangeable functions. Why? DBE is parametrised by probabilities taken from the elementary population level, but EBE takes probabilities form the holistic level of the vegetation stand itself. Seamless adaption of concepts and terminology in stand-level community ecology permeates the entire text. Key references are listed and a complete step-by-step example included in the Appendices.

## A CONVERSATION

What are you doing? She asked. I am negating *argumentum ad populum*, I said. It goes like this:

P1: X is popular
P2: I should not do what is popular
P3: I should not do X

Why are you doing that? She insisted. Simple, I said. I am not a bandwagon man.

# DIVERSITY ANALYSIS, HOLISTIC ENERGETICS, AND STATISTICS

The resonator complex model of the vegetation stand

László Orlóci FRSC

Ecologia Quantitativa, UFRGS, Porto Alegre, Brazil

SCADA Publishing – Canada 2015

**Refer to this monograph:**

Orlóci, L. 2015. Diversity analysis, holistic energetics, and statistics. The resonator complex model of the vegetation stand. SCADA Publishing, Canada. Online Edition: https://createspace.com/5783923

**Look for these monographs:**

Orlóci, L. 2015. Energy-based vegetation mapping. A case study in statistical quantum ecology. SCADA Publishing, Canada. Online Edition: https://createspace.com/5495773

Orlóci, L. 2014. The vegetation process. A holistic study of long-term community energetics in East Beringia. SCADA Publishing, Canada. Online Edition: https://createspace.com/4760258

Orlóci, L. 2013. Quantum analysis of primary succession. The energy structure of a vegetation chronosere in Hawai'i Volcanoes National Park. SCADA Publishing, Canada. Online Edition: https://createspace.com/4452597

Orlóci, L. 2014. Quantum Ecology. Energy structure and its analysis. SCADA Publishing, Canada. Online Edition: https://createspace.com/4406077

Orlóci, L. 2013. On the Energy Structure of Natural vegetation. In search for community governance rules. SCADA Publishing, Canada. Enlarged Online Edition: https://createspace.com/4153484

Orlóci, L. 2012. Self-organisation and Mediated Transience in Plant Communities. SCADA Publishing, Canada. Enlarged Online Edition: https://createspace.com/3585127

Orlóci, L. 2014. Statistical Ecology. The quantitative exploration of nature to reveal the unexpected. SCADA Publishing, Canada. Online Edition: https://createspace.com/3476529

Orlóci, L. 2012. Statistical multiscaling in dynamic ecology. Probing the long-term vegetation process for patterns of parameter oscillations. SCADA Publishing, Canada. Online Edition: https://createspace.com/3830594

Orlóci. L. 2011. Problem flexible computing in statistical ecology. SCADA Publishing, Canada. Online Edition: https://createspace.com/3574792

ISBN-13: 9781517687069   ISBN-10: 1517687063
V 2016-05-07

Find further information:
https://sites.google.com/site/statisticalecology/

Address for all communications: lorloci@uwo.ca

Diversity analysis, holistic energetics and statistics

# Table of contents

| | |
|---|---|
| Table of contents | 5 |
| Preliminaries | 6 |
| Operational vegetation unit | 6 |
| Resonator complex model | 6 |
| Characteristic traits | 6 |
| A stand's relevé | 7 |
| DBE centred diversity analysis | 7 |
| EBE centred energetics | 8 |
| More on the EBE function | 9 |
| Metaphor issue | 10 |
| Regularity conditions defined | 12 |
| Status of taxa | 13 |
| Three-parted EBE structure | 14 |
| Untangling a perennial confusion | 14 |
| Statistics: a family of dialects | 16 |
| Summary | 17 |
| Reference bibliography | 19 |
| Index | 21 |
| Appendices | 23 |
| A1. Artificial data set | 23 |
| A2. EBE and related statistics | 25 |
| A3. Emergent (ghost) EBE | 26 |
| A4. Decomposition of nH in (A+B+C) by hierarchical level | 27 |
| A5. $E_{Env}$ term | 28 |
| A6. E equation unfolded | 29 |
| Supplementary information | 31 |
| Reader's notes | 41 |
| Biographic notes … | 43 |

# Preliminaries

### Operational vegetation unit

The very nature of the topic, as foretold in the main title, mandates consideration of both theoretical and utilitarian aspects. With the needs of community ecology in mind, I have to use *operational vegetation units*. My term for these is the *vegetation stand,* a plant community within *arbitrary* areal limits. Delineation is the gestalt or formal analytical criteria regarding homogeneity.

### Resonator complex model

The technical presentation of disorder-based entropy (DBE) and energy-based entropy (EBE) is conveniently done in the context of the resonator complex model of Max Planck (1901). Translated into community ecology's terms, the complex is equated to the unit plant community, in our case the vegetation stand. Resonators are identified as plant populations, the taxa which compose the vegetation stand.

### Characteristic traits

The unit plant community's characteristics traits are population and stand level. Species frequency is a typical population level trait. The stand total of all species frequencies is the corresponding stand level trait. I use these in the example, but it should not imply more than an indication that population performance has to be measured in countable units.

### A stand's relevé

The frequency record of a vegetation stand is presented in the manner of an n-valued, T-totalled vector $F = [f_1 \; f_2 \; ... \; f_n]$. **F** is called *relevé* in French and *Aufnamen* in German. Symbol n stands for the number of distinct populations called *taxa*. I use the term 'taxa' not to commit the analysis to a single system of plant classification *a priori*. For example, I could say 'species population' if I intended to emphasise the population's uniformity in terms of inheritance. But I can recognise populations for their uniform plant functionality, such as life form, in which case the taxon becomes a 'functional type'. In my approach, the taxon type is not fixed until the study plans are completed.

To give a first insight into the difference between DBE and EBE, it is sufficient to point out that DBE is parametrised by probabilities defined for the elements of **F**, different from EBE which is parametrised by a probability which is a function of the stand totals n and $T = \sum_{i=1}^{n} f_i$ . Because of this reason, I refer to the EBE function as *holistic scalar*, and to the EBE-based approach as *holistic energetics.*

# DBE centred diversity analysis

DBE is the basis of the most broadly used type of diversity analysis. Disorder is measurable within **F**. It is maximal when **F** is an equidistribution, and minimal when **F** is most contagious.

The ecological perimeters of diversity analysis are treated in texts on quantitative ecology (e.g., Pielou, 1977). My preferred references for underlying theory include Shannon (1948), Kullback (1959), Rényi (1961) and Brillouin (1962). They create context on the specifics presented in a later section, in which I compare DBE to EBE at length.

# EBE centred energetics

My EBE is Max Planck's (1901) energy-based entropy function without the constants:

$$E = nH = -\ln P = \ln C$$

In this, $nH$ is a single symbol, $P = \dfrac{1}{C}$, and C stands for the combinatorial,

$$C = \frac{(T+n-1)!}{T!(n-1)!} \approx \frac{(T+n)^{T+n}}{T^T n^n}$$

The last expression in the above is Stirling's approximation, used by Max Planck. It can be seen why I say that the EBE function is parameterised at stand level by $n$ and $T$.

The analytical tools of stand level energetics target the decomposition of EBE in the manner of

$$E = E_{Phy} + E_{Env} + E_{Rnd}$$

The first two terms on the right hand side are measurable. $E_{Phy}$ is proxy for the amount of energy spent by the phylogenetic process to produce the current level of plant taxon richness. The concomitant $E_{Env}$ is an estimate of energy spent in the process of recent environmental mediation which sorted the plant populations into distinct vegetation stands. The third term, $E_{Env}$, is a residual in the manner of the difference,

$$E_{Rnd} = E - E_{Phy} - E_{Env}$$

As a provisional decision, not uncommon in statistical practice, the unidentified effects, which are responsible for $E_{Rnd}$, are assigned to the grab-bag of the so-called error generating random events. The 'error' designation notwithstanding, dependent on the context of the problem, $E_{Rnd}$ can be an emergent or mutual EBE attributed to *interaction*.

# More on the EBE function

I have introduced already the EBE concept. It invites the question, why I use $E = - \ln P$ and why not $S_N = k \log W +$ constant ? The original function has constants. Will the analysis suffer if I leave out the constants? Before I give the answer, I have to introduce further symbols:

$S_N$ -- Max Planck's symbol for the energy-based entropy (potential energy) state of his N-resonator complex.

ln -- natural logarithm.

k -- Max Planck's universal constant which applies when the complex is placed into a homogeneous radiation field.

$W = \dfrac{1}{P}$ -- is C.

'constant' – an arbitrary number which makes $S_N$ positive.

I already defined our EBE function $E = -\ln P$ without the constants. So then what are the consequences of leaving out the constants? I refer to an earlier explanation (Orlóci 2015) which I now paraphrase:

   a. If the second constant compensates for the lack of a negative sign in

$$S_N = k \log W + \text{constant}$$

when 'constant' is removed, the expression has to be written

$$S_N = -k \log W$$

b. Constant k is not trivial in the original context. But I do not see why it should be needed for what I do with the E function in community ecology. My objective is a comparison of stands on a relative basis, and measurement of the relationship of E and specific forcing factors taken in an ecological context. By setting k equal to -1 neither of these objectives should be short changed.

# Metaphor issue

This came up in questions regarding the validity the use of 'energy' in connection with E. The question is frivolous, but need discussion to liberate the application of EBE from a misconception. This is what I should say:

1) Max Plank marshals the proof, after assuming specific regularity conditions, under which EBE is in fact an alternative scalar of the potential energy level in his resonator complex.

2) The suggestion that 'energy' in EBE is nothing more than a metaphor, and 'energy' when used in calorie measurements is not a metaphor, is simply a mute issue. To see this, one should consider the fact that energy has never been defined (Feynman 1964) and for that very reason it cannot be measured. As a matter of fact, all energy scalars, no matter in what units, joules or otherwise, always measure a manifestation of spent energy. This being the case, it must be true that there are as many possible alternative energy scalars, and units of measurements, as there are possible measurable manifestations of spent energy.

3) I consider this further. The rise of ambient temperature in the vicinity of a heat source, is a manifestation of spent energy which is source for the generated heat causing the rise. When one expresses it as calorie in joules, 'calorie' is the alternative scalar of spent energy. When I describe the potential energy level in a resonator complex, I find the manifestations of energy in the sum $T$. With $n$ resonators in the complex, the EBE equation $E = \ln C$ is the alternative scalar for potential energy level of the unit community measured in nats. 'Energy' in nuts or in joules is in neither case or in both cases is a 'metaphor'. The point is that scalar functions represent alternative ways of measuring energy on proxy scales.

# Regularity conditions defined

The problem with energy-based entropy is not nomenclatural. It is much deeper. Before E is applied, it is required to show that certain regularity conditions exist in the unit plant community. What are these conditions?

1) Observe that since P is a probability, which is universal, and E is a function of P, there is, associated with any value of P a single E for C distinct complexes of the same resonator set. The smaller the value of P, the greater is E.

2) Observe further that any $f_i$ in **F** is regarded as a chance value. As chance events go, the highest probability event is the mean itself. For that reason I prefer to regard any $f_i$ as a best representation of the mean. What do I commit myself to by saying this? I have to assume the total rule of chance in the assembly process of the unit plant community.

3) Since I define the unit plant community as an area unit of the landscape, not greater than the phytosociologist's sample plot - whose exact size depends on sampling consideration, no matter where the sample plot is laid out on the ground - the rule of chance in stand assembly must be true. This is a requirement, but can it be that way? I believe not just that it can be, but it will, if the spatial pattern of vegetation and environment, individually and jointly, is random within the sample plot, and if the stand is homogeneous. How do I test for these? Pattern analytical techniques are available for testing the ground pattern's actual randomness (Greig-Smith 1982). I prefer an adaptation of our method with ecologists Valério DE Patta Pillar and Otto Wildi (Orlóci and Pillar 1989, Orlóci and Wildi

1987) to determine the critical sample plot size within which the target structure of the stand has firm stability. Random pattern and stable structure adds up to homogeneity. The phytosociologist's gestalt in sampling unit selection is a good starting point in the tests (Orlóci 1993a,b).

4)  Each element $f_i$ in **F** is assumed to be a count of energy units (not the amount of energy). I regard $f_i$ a value handed to us by chance from the C-valued population of complexes. By virtue of the fact that under the rule of chance the average has the highest probability of being chosen, $f_i$ is a best estimate of that mean.

5) As I noted earlier, E as defined do not depend on the level of disorder in the observed **F**. All should accept that the observed disorder in **F** cannot be responsible for the potential energy level of **F**. Yet there is no energy without disorder. The disorder connection of E is intrinsic in the convoluted process of phylogeny and environmental sorting.

# Status of taxa

E is parameterised by the actual number of taxa (n) and the sum of taxon performances (T). 'Actual' puts emphasis on the context of the operations which can pool or split taxa, or interchange their contents to create new taxa. A simple example of this is the passing from a species-based taxonomy to function based taxonomy (Orlóci 2015 and references therein). The analysis in this respect is completely dialectic and requires keeping in mind that in the quantity $H = \dfrac{E}{n}$ one always use the actual n. I refer to H

as the one-taxon, one resonator, or average energy-based entropy.

## Three-parted EBE structure

The EBE level in a vegetation stand is a sum of three independent components which are additive,

$$E = E_{Phy} + E_{Env} + E_{Rnd}$$

$E_{Rnd}$ is a residual in the manner of the difference,

$$E_{Rnd} = E - E_{Phy} - E_{Env}$$

Following statistical ecology's practice, I assign the unidentified effects responsible for $E_{Rnd}$ to the grab-bag called error generating events. I may call it the emergent nH, interaction nH, or shared nH, depending on the context of the operations.

## Untangling a perennial confusion

I already have drawn an unbridgeable demarcation line in practice between DBE and EBE. Now, I will argue why the family of logarithmic expressions, which is the sum of negative p ln p terms, does not include nH = ln C or

$H = \dfrac{E}{n}$. In other words, the field, defined in the works of Shannon (1948),

Kullback (1959), Rényi (1961), and Brillouin (1962), exploited in diversity analysis and in information analysis (Orlóci 1991, 2014), does not include Planck's (1901) energy-based entropy.

For reference, the following is give:

$$H_\alpha = \frac{1}{1-\alpha} \ln \sum_{i=1}^{s} p^\alpha \quad \text{-- Rényi's generalised entropy of order } \alpha.$$

$$I_\alpha = \frac{1}{\alpha-1} \ln \sum_{i=1}^{s} \frac{p^\alpha}{q^{\alpha-1}} \quad \text{-- Rényi's generalised information of order } \alpha.$$

$$H = -\sum_{i=1}^{s} p_i \ln p_i \quad \text{-- Shannon's average entropy.}$$

Note that Shannon's entropy is the same as Rényi's entropy of order one with order parameter $\alpha$ approaching 1. In practice this condition can be fulfilled by a very pedestrian solution: select a value for $\alpha$ close to 1, say $\alpha = 0.999999$.

$$2I = 2\sum_{i=1}^{s} f_i \ln \frac{p_i}{q_i} \quad \text{-- Kullback's discrimination information statistic.}$$

This is the same as twice Rényi's generalised information when the order parameter $\alpha$ is approaching 1. The Rényi and Kullback equations are parametrised by the same p and q. Letter q stands for the random expectation of p. The definition of s is consistent with its use it in the equation,

$$p_1 + p_2 + ... + p_s = q_1 + q_2 + ... + q_s = 1$$

In other words letter s stands for the same as n or N.

It should be clearly seen now that $E = \ln C$ is neither Rényi's generalised entropy nor his generalised information. It should be seen too that the similarity of $E = \ln C$ in form and Brillouin's (1962) information function $I = \log_2 C$ is superficial. To verify this, readers only need to compare the definition of Brillouin's $C$,

$$C = \frac{N!}{f_1! \, f_2! \, \dots \, f_s!}$$

and Max Planck's $C$,

$$C = \frac{(T + n - 1)!}{T!(n - 1)!}$$

The capital letter $N$ and lowercase $n$ in the above expressions represent the same thing.

Note that Brillouin's $f_i$ is an element in the frequency vector **F** from which Rényi's $p_i$ is derived, such as $p_i = \frac{f_i}{T}$. Note further, that $I = \log_2 2$ is one bit of information and Brillouin's $I$ has maximum $\log_2 s$ bits. Useful to know further that when the $f_i$ numbers are large, say 100 each or greater, in terms of Brillouin's $I$ the quantity $\dfrac{I/s}{\log_2 e}$ will come close in value to Shannon's entropy which is the same as Rényi's entropy of order one.

# Statistics: a family of dialects

Considering the novelty of EBE related data analysis in community ecology, a brief discussion is in order regarding Statistical Quantum Ecology which supplies the analytical tools equivalent in a statistical dialect. I explain.

My first approximation the science 'Statistics' is rather pragmatically motivated. As such, to me Statistics is nothing more in principle than the coupling of probability theory with a characteristic function drawn from physical science. I distinguish and talk about statistical dialects in this sense.

For example, when probability theory is coupled with the system of moments and product moments, foundation is laid for what I call the 'Fisherian' statistical dialect (Orlóci 1991, 1993a, 2001). This is the statistical dialect most commonly used by ecologists. When the coupling involves probability theory and Rényi's (1961) information divergence of order one, the foundation of Kullback's (1959) information divergence based statistics is defined. Kullback's is yet another statistical dialect.

Continuing in this wain, I see only practical limits to the possible number of statistical dialects that can be construed. Statistical Quantum Ecology is one of them. Its foundations involve a coupling of probability theory with EBE from physics (Orlóci 2013 a,b, 2014a,b, 2015).

# Summary

It should be clear from what I have said so far that  DBE and EBE are different 'species' in the family of logarithmic functions of probability. I observe once more that parameterisation of $nH = - \ln P$, unlike the -p ln p and $p \ln \dfrac{p}{q}$ type expressions, does not reach down to the individual taxon's p and q values, but uses the stand level P related to T and n of **F**. The **EBE** value will come out to be the same value in all C permutations of **F**. This is not so with DBE which can have different values from some minimum not quite zero to ln n.

EBE is a function of probability, which makes it easy to convert E back into probability simply by negative exponentiation. Therefore, EBE is conveniently used in tests of statistical hypotheses. Another property of EBE is its additivity. This property allows statistical isolation of nH components into independent terms in the manner of $E = E_{Phy} + E_{Env} + E_{Rnd}$ . It must not be forgotten that the elements represent equitability and mutuality as in a Venn diagram (Orlóci 2015). The example refers to this property. The computational tasks with nH are relatively simple and can be executed on a spread sheet. I include Appendices to show this.

The above discussion sends a message to the user: do not try to interpret E entirely from what you know about the individual taxa. The *reductionist ignoramus* would do such a thing, and in all likeliness the model would not fit reality. A model of community energetics should begin formulated at the stand level with EBE, then enriched thereafter by expertly incorporating whatever information is available about the community elements.

# Reference bibliography

Brillouin, L. 1962. Science and information theory. 2nd ed. Academic Press, New York.

Feynman, R. 1964. The Feynman Lectures on Physics. Vol. 1. U.S.A., Addison Wesley. ISBN 0-201-02115-3.

Greig-Smith. P. 1983. Quantitative plant ecology. 3$^{rd}$ ed. Blackwell Scientific, London.

Kullback, S.M. 1959. Information theory and statistics. Dover, New York.

Orlóci, L. 1991a. Poorean approximation and Fisherian inference in bioenvironmental analysis. Research Trends. Advances in Ecology 1:65-71.

Orlóci, L. 1991. Entropy and Information. Ecological Computations Series: Vol. 3. SPB Academic Publishing, The Hague.

Orlóci, L. 1993a. Conjectures and scenarios in recovery study. Coenoses 8:141-148.

Orlóci, L. 1993b. The complexities and scenarios of ecosystem analysis. In: G. P. Patil and C. R. Rao, Multivariate Environmental Statistics, pp.421-430, North Holland/Elsevier, New York.

Orlóci, L. 2001. Prospects and expectations: reflections on a science in change. Community Ecology 2: 187-196.

Orlóci, L. 2014a. Statistical Ecology. The quantitative exploration of nature to reveal the unexpected. SCADA Publishing, Canada. Online edition: https://createspace.com/3476529

Orlóci, L. 2014b. The vegetation process. A holistic study of long-term community energetics in East Beringia. SCADA Publishing, Canada. Online edition: https://createspace.com/4760258

Orlóci, L. 2013a. Quantum Ecology. Energy structure and its analysis. SCADA Publishing, Canada. Online edition: https://createspace.com/4406077

Orlóci, L. 2013b. Quantum analysis of primary succession. The energy structure of a vegetation chronosere in Hawai'i Volcanoes National Park.

SCADA Publishing, Canada. Online edition:
https://createspace.com/4452597

Orlóci, L. 2015. Energy-based vegetation mapping. A case study in statistical quantum ecology. SCADA Publishing, Canada. Online Edition:
https://createspace.com/5495773

Orlóci, L. and V. De Patta Pillar. 1989. On sample size optimality in ecosystem survey. Biométrie – Praximétrie 29:173-184.

Orlóci, L. and V. De Patta Pillar. 1991. On sample size optimality in ecosystem survey. Reprinted in: Feoli, E. and L. Orlóci (eds.), Computer Assisted Vegetation Analysis, pp. 41-46. Kluwer Academic Publishers, London.

Pielou, E.C. 1977. Mathematical ecology. Wiley-Interscience, New York.

Planck, Max. 1901. On the law of distribution of energy in the normal spectrum. Annalen der Physik Vol. 4, p. 553 et seq. -- To download this paper, go to https://sites.google.com/site/statisticalecology/ then click item 40 in the Selected References section on Laszlo's side.

Podani, J. 2015. A Növények evoluciója és osztályozása. (Evolution and classification of plants.) ELTE Eötvös Kiadó, Budapest.

Rényi, A. 1961. On measures of entropy and information. In: J. Neyman (ed.), Proceedings of the 4th Berkeley Symposium on Mathematical Statistics and Probability, pp. 547-561. University of California Press, Berkeley.

Shannon, C.E. 1948. A mathematical theory of communication. Bell System Tech. J. 27: 379-423.

Wildi, O. and L. Orlóci. 1987. Flexible gradient analysis: a note on ideas and an application. Coenoses 2: 15-19.

Wildi, O. and L. Orlóci. 2007. Essay on the study of the vegetation process. In: F. Kienast, O. Wildi and S. Ghosh (eds.), A Changing World. Challenges for Landscape Research. pp. 195-207.

# Index

actual n, 14
alternative scalar, 11
amount of energy, 8, 13
arbitrary areal limits, 6
Beringia, 4, 19
Bibliographic notes ..., 43
Brillouin, 8, 14, 16, 19
Brillouin's expression, 16
chance, 26
chronosere, 4, 19
classification, 25
comparison of stands, 10
complex, 6, 9, 11, 26
constants, 9
context, 6, 10, 13
context of the operations, 13
critical sample plot size, 13
DBE, 7, 14, 17
disorder, 13
disorder connection of E, 13
divergence times, 25
EBE, 6, 7, 8, 9, 10, 11, 14, 16, 17, 18, 26, 27, 28
ecology, 4, 7, 14, 16, 18, 19, 20, 43
energetics, 1, 3, 4, 8, 12, 17, 19
energy, 4, 9, 10, 11, 13, 14, 19, 20, 26, See
energy structure, 4, 19, 26
energy units, 13
energy-based entropy, 9, 14, 15
entropy, 8, 16, 20
environmental mediation, 9, 13
error, 9
error generating events, 14

evolution, 25
expectation, 15
family, 43
flipping, 26
forcing factors, 10
frequency vector, 7, 16
functional plant types, 13
generalised entropy, 16
ghost states, 26
Greig-Smith, 12, 19
hierarchical relevé, 25
holistic energetics., 1, 7
holistic scalar, 7, See
homogeneity, 11
Homogeneity, 12
information, 4, 16, 20
instability, 26
Kullback, 8, 14, 15, 17, 19
level of disorder, 13
manifestations of energy, 11
Márta Mihály, 43
Max Planck, 8, 9
Max Plank, 11
metacommunity, 27, 28
metaphor, 10
model, 6
multiscaling, 4
new taxa, 13
node, 24
number of taxa, 13
observed disorder, 13
operational vegetation units, 6
order parameter $\alpha$, 15

Orlóci, 3, 4, 10, 12, 15, 17, 19, 20, 43
phylogenetic principles, 25
phylogenetic process, 8
phylogeny, 13
phytosociologist's gestalt, 13
Pielou, 8
Pillar, 12, 20
Planck's expression, 16
Podani, 25
population, 6
probability, 12, 17, 18, 26, 28
probability theory, 17
process, 4, 9, 13, 19, 20
Quantum analysis, 4, 19, See
quantum ecology, 4, 20
Quantum Ecology, 4, 19
random events, 9
regularity conditions, 12
Rényi, 8, 14, 15, 16, 17, 20
residual, 9, 14
resonator, 6, 9, 11, 12
resonator complex, 6, 11
richness, 9
sample plot, 13
scalar functions, 11, 17

Shannon, 8, 14, 15, 16, 20
spatial pattern, 12
species, 13, 17, 23, 24, 26
split taxa, 13
stability, 13, 26
stand, 6, 8, 26
statistical dialects, 17
Stirling's approximation, 8
symbols, 9
systematic status, 23
taxa, 13, 23, 24
taxon performances, 13
taxon ranking, 25
the energy level, 14
trajectory, 43
transitions, 43
UFRGS, 3
unit plant community, 6
universal constant, 9
vegetation, 4, 6, 9, 12, 14, 19, 20, 23, 26, 43
vegetation process, 4
vegetation stand, 6, 14, 23
vegetation structure, 13
Wildi, 12, 20

# Appendices

## A1. Artificial data set

The symbolic representation of a stands description is

$$\mathbf{F} = \begin{bmatrix} f_1 & f_2 & \dots & f_n \end{bmatrix}$$

I refer to **F** as a relevé of the vegetation stand. The elements of **F** are counts for n taxa. **F** is amended with a code matrix,

$$\mathbf{U} = \begin{bmatrix} u_{11} & u_{12} & \dots & u_{1s} \\ u_{21} & u_{22} & \dots & u_{2s} \\ \dots & \dots & \dots & \dots \\ u_{n1} & u_{n2} & \dots & u_{ns} \end{bmatrix}$$

The elements in **U** specify the systematic status of the taxa. For example, $\mathbf{U}_1 = \begin{bmatrix} 1 & 1 & 1 & 1 \end{bmatrix}$ is telling us that taxon 1, if a species, belongs to Class 1, Order 1, Family 1 and Genus 1.

**F** and **U** together define a hierarchical relevé. There are three such relevés included in Figure A1. Six species and two active hierarchical levels are involved above the species. In principle any number of levels are allowed with any number of nodes (taxa). The numerical **F** and **U** matrices are in Table A1 and A2.

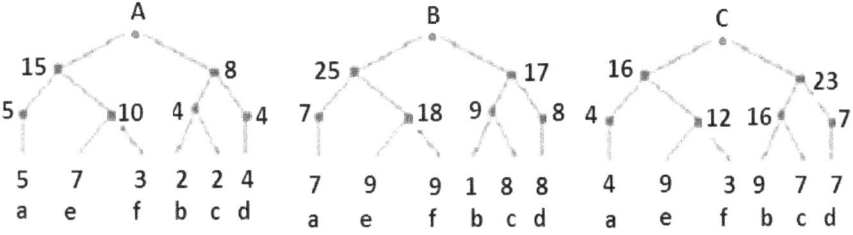

Figure A1. Hierarchical relevés for six species and two  hierarchical levels above species. First (top) level: two Families. Second level: four Genera.

Table A1. Frequency data for three stands and six species corresponding to Figure A1.

| F vectors | a | e | f | b | c | d | T | n |
|---|---|---|---|---|---|---|---|---|
| A | 5 | 7 | 3 | 2 | 2 | 4 | 23 | 6 |
| B | 1 | 9 | 9 | 1 | 8 | 8 | 36 | 6 |
| C | 4 | 9 | 3 | 9 | 7 | 7 | 39 | 6 |
| Total | 10 | 25 | 15 | 12 | 17 | 19 | 98 | 18 |
| Total B+C | | | | | | | 75 | 12 |

Table A2. Code matrix **U** for the six species in Figure A1.

| Species | Family | Genus |
|---|---|---|
| a | 1 | 1 |
| e | 1 | 2 |
| f | 1 | 2 |
| b | 2 | 3 |
| c | 2 | 3 |
| d | 2 | 4 |

The graphs constructed in Figure A1 are informative, but need not be drawn. Matrices **F** and **U** are sufficient for input in all calculations to be performed. The cumulants (5, 10, 4, 4, etc.) are needed in advanced operations. Note how the parameter n changes value through the hierarchical levels from 2 to 4 to 6.

When I use any type of taxa, it is implicit that I am dealing with unique populations of plants recognised by shared characteristics. In Figure 1A the null level taxa are species populations, distinct by inheritance. Any node in the dendrogram identifies a point at which a higher level taxon are split into lower level taxa in the course of evolution. The hierarchical relevé for species is indeed assumed to be a local proxy for the phylogenetic tree.[1]

---

[1] To answer arguments in specific cases is beyond my intention. I leave that discussion to the experts of evolutionary plant systematics. I was kindly warned that ranking taxa by their systematic status as in Figure 3 is quite

## A2. EBE and related statistics

I begin with general analysis in which I introduce different EBE and EBE-related quantities:

| Relevé | T | n | nH | H | P | 1-p |
|--------|-----|-----|--------|-------|-------|-------|
| A | 23 | 6 | 14.785 | 2.464 | 0.085 | 0.915 |
| B | 36 | 6 | 17.225 | 2.871 | 0.057 | 0.943 |
| C | 39 | 6 | 17.670 | 2.945 | 0.053 | 0.947 |
| A+B+C | 98 | 18 | 50.063 | 2.781 | 0.062 | 0.938 |
| B+C | 75 | 12 | 34.904 | 2.909 | 0.055 | 0.945 |

| $W_{AB}$ | $\omega_{AB}$ | $\overline{m\omega}_{AB}$ | $m\omega_{AB}$ |
|-------|-------|-------|-------|
| 0.156 | 0.558 | 0.816 | 0.583 |
| 0.107 | 0.462 | 0.621 | 0.771 |
| 0.100 | 0.446 | 0.591 | 0.806 |
| 0.116 | 0.482 | 0.658 | 0.729 |
| 0.103 | 0.454 | 0.606 | 0.789 |

Definitions:

---

arbitrary, just as much, as I see it, as any applicable plant systematics upon which the available field manuals are bases, and from which I construct my taxon ranking. This, notwithstanding the designation of these manuals as being based on phylogenetic principles, is beyond my intent to endorse or reject. I wait with radical changes of my hierarchical relevé scheme until I am handed a plant system based on pairwise divergence times of taxa which covers the local flora. I chance saying that the changes may not happen soon, despite the earth shaking changes in works which are making classical plant systematics appear increasingly obsolete. I see this happening from János Podani's seminal book "The evolution and classification of plants" published in Magyar by Elte Eötvös Kiadó, Budapest, 215.

$nH = -\ln P = (T+n)\ln(T+n) - T\ln T - n\ln n$ -- EBE level in the stand. T and n are stand totals (see A1). Note, "ln" stands for the natural logarithm. Herefore, E is measured in natural units (nat).

$H = \dfrac{nH}{n}$ -- One-taxon EBE.

$P = e^{H}$ -- Probability associated with H.

$w_{AB} = 1 - P_A^2 - P_B^2$, $P_A = \dfrac{1}{C}$, $P_B = 1 - P_A$ -- Instability level in the stand's EBE structure. The values of $w_{AB}$ range from 0 (complete stability) to 0.5 (complete instability). This is easy to see if 1 or 0.5 is substituted for P. How do I interpret $w_{AB}$? There is more than one way to do this. For example, as $w_{AB}$ increases, the chance of the stand's energy structure flipping by pure chance into one of its C-1 ghost states increases. The $w_{AB}$ is in squared probability units.

$\omega_{AB} = \sqrt{2w_{AB}}$ -- a probability.

$^{-}m\omega_{AB} = -\ln(1 - \omega_{AB})$ -- the *unit instability moment*. This is the strength of instability or equivalently the unit linear moment forcing the stand's energy structure to flip into a ghost state.

$m\omega_{AB}$ -- the *unit stability moment*.

### A3. Emergent (ghost) EBE

I assume a sere (catena) of vegetation stands, linked by some natural trait, such as succession in a chronosere, orographic effect on landscape catena, etc. I designate the three relevés (A, B. C), or when pooled, by symbol (A+B+C). The csse to be considered is the single relevé (A) which has to join (A+B) to complete the catena (A+B+C). Consider the nH values in Table A3 taken from section A2.

Table A3. Determination of the emergent EBE when (A) is joined to (B+C).

|  | nH(A+B+C) | nH(B+C) | nH(A) | dnH(A) | gnH(A) |
|---|---|---|---|---|---|
| Catena | 50.062958 | 34.90352 | 14.78466 | 15.15944 | 0.374784 |

Definitions:

$dnH(A) = nH(A + B + C) - nH(B + C)$ -- nH divergence.

$gnH(A) = dnH(A) - nH(A)$ -- the ghost nH. This is the size of the emergent EBE when (A) is joined to (B+C).

## A4. Decomposition of nH in (A+B+C) by hierarchical level

This can be done on each dendrogram. I do it for the dendrogram of the metacommunity formed by pooling the three relevés. I note that the species complements are identical in the three relevés.

| Metacommunity | Level | T | n | nH | dnH | dn | Specific dH | P |
|---|---|---|---|---|---|---|---|---|
|  | 1 | 98 | 2 | 9.804 | 9.804 | 2 | 4.902 | 0.007 |
|  | 2 | 98 | 4 | 16.875 | 7.071 | 2 | 3.536 | 0.029 |
|  | 0 | 98 | 6 | 22.939 | 6.064 | 2 | 3.032 | 0.048 |
|  |  |  |  | Total | 22.939 | 6 |  |  |

Interpretation:

The T value comes from Table A1. The n values count the dendrogram nodes on different levels. The arithmetic is self-explanatory.

Two alternative arrangements could be used:

1) Disregard species identities and join the dendrograms sequentially.

| Level | T | n | nH | dnH | dn | Specific nH | P |
|---|---|---|---|---|---|---|---|
| 1 | 98 | 6 | 22.939 | 22.939 | 2 | 11.470 | 0.000 |
| 2 | 98 | 12 | 37.907 | 14.968 | 2 | 7.484 | 0.001 |
| 0 | 98 | 18 | 50.063 | 12.156 | 2 | 6.078 | 0.002 |
| | | | | 50.063 | 18 | | |

2) When the species lists of the relevés only have partial overlap, a new dendrogram is created for the metacommunity and analysed as such.

How do I interpret the numerics? I start with the dnH values. These are the EBE quantities specific to the steps from level to level. The sum of dnH is the nH value on level 0. Note, dnH declines form level 1 down. In a real case this could give us guidance which systematic level I should take as the base level for data recording. The specific dH is the single-node or average EBE value of the level. The P values tell me that I am dealing with reasonably low probability events which warrant to consider each dH value statistically significant.

## A5. E_Env term

In a real case it could happen that the sere (A, B, C, …) is concomitant to directional environmental forcing on the sere. Assuming in the example that environmental forcing has 3 levels of intensity, then there is:

| | T | n | nH | H | P |
|---|---|---|---|---|---|
| Environment | 98 | 3 | 13.505 | 4.502 | 0.011 |
| | 33 | 3 | 10.326 | 3.442 | 0.032 |

The second row value T=33 is an average taken to the nearest integer. Considering that P is rather small, I regard the H values statistically significant.

### A6. E equation unfolded

This equation $E = E_{Phy} + E_{Env} + E_{Rnd}$ is interpreted according to the Venn diagram,

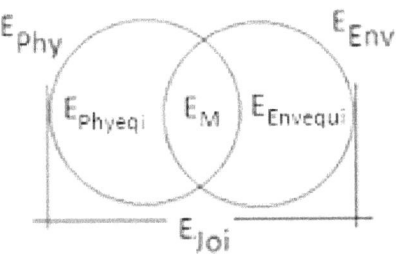

Legend: equi – equivocation; M – mutual, shared or emergent; joi – joint.

The corresponding numerical values are:

|        | T  | n   | nH (Venn) | %      | nH total | H      | P      |
|--------|----|-----|-----------|--------|----------|--------|--------|
| EPhy   | 98 | 6   | 22.9393   | 45.82  | 36.5584  | 3.8232 | 0.0219 |
| EEnv   | 98 | 3   | 13.5045   | 26.98  | 27.1237  | 4.5015 | 0.0111 |
| ERND=EM| 98 | *3  | 13.6192   | 27.20  |          | 4.5397 | 0.0107 |
| E=EJoi | 98 | 18  | 50.0630   | 100.00 |          |        |        |
| Total  |    |     |           |        | 63.6821  |        |        |

| *Iteration | 98 | 4 | 16.8752 |
|------------|----|---|---------|
|            | 98 | 3 | 13.5045 |
|            | 98 | 2 | 9.8039  |

Conclusion:

In a real case I could say that the $E_{Phy}$ component is overwhelming, $E_{Env}$ is strong, and so is the $E_{Rnd}$ component which I attribute to an emergent effect. All H values in the above table are statistically significant.

A further note is in order. It can be shown that with the difference $|n_{Phy} - n_{Env}|$ decreasing, the size of EM increases and the energy-based entropy structure's stability declines.

# Supplementary information

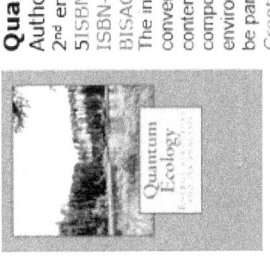

**Quantum Ecology:** Energy structure and its analysis
Authored by Dr Laszlo Orloci FRSC
2nd enlarged edition: 201
5ISBN-13: 978-1517432935 (CreateSpace Assigned)
ISBN-10: 1517432936
BISAC: Science / Life Sciences / Ecology
The infusion of quantum theoretical principles allows the study focus of ecological energetics to shift from the conventional calorific (trophic) flow in ecosystems to the potential energy structure of the vegetation. The books contents cover the theory and techniques in a unique account centred on the energy equation. The equation's component terms define energy footprints specific to ecology's basic processes, such as historic phylogeny, current environmental mediation of transience, and chance. What gives practical value to the energy equation is its ability to be parameterised by the usual type of survey or experimental data
CreateSpace eStore:  https://www.createspace.com/5750582

# Energy-based vegetation mapping  [f Like] ⟨0⟩

## A case study in statistical quantum ecology.

**Authored by Dr Laszlo Orloci**

The book presents further details of a new paradigm intended for statistical studies of stand-level vegetation energetics. The approach is holistic and the techniques are quantum ecological. The case study's objective is stand-level vegetation mapping by energy criteria. Since the vegetation process is conceived as an energy structural phenomenon, the obvious choice for mapping is an energy structural component.

The book's classification model assumes a three-parted structure with components issuing from historic phylogeny, current environmental mediation, and the ubiquitous random events. The modus operandi is simple: isolate the structural components analytically, construct the maps for display on the sampling grid, then probe the maps and other numeric results for generalizable regularities. Regarding the data base, most types of vegetation survey data are admissible.

List Price: $15.00

[Add to Cart]

| | |
|---|---|
| **Publication Date:** | May 12 2015 |
| **ISBN/EAN13:** | 1512180769 / 9781512180763 |
| **Page Count:** | 92 |
| **Binding Type:** | US Trade Paper |
| **Trim Size:** | 6" x 9" |
| **Language:** | English |
| **Color:** | Black and White |
| **Related Categories:** | Science / Life Sciences / Ecology |

Order from: https://createspace.com/5495773

**THE VEGETATION PROCESS: A holistic study of long-term community energetics in East Beringia**
Authored by Dr Laszlo Orlóci

6" x 9" (15.24 x 22.86 cm)

Black & White on White paper

216 pages

ISBN-13: 978-1499142068 (CreateSpace-Assigned)

ISBN-10: 1499142064

BISAC: Science / Life Sciences / Ecology

ORDER FROM CREATESPACE E-STORE:

https://www.createspace.com/4760258

Process, as the Book uses this term, implies simultaneous execution of two fundamental functions in continuity. One creates complexity, the other reduces it. Ecologists refer to these as community assembly and disassembly. The process requires energy input which determines the momentary potential energy state of the community. This is measurably true in terms of Max Planck's energy-based entropy. We find potential energy increasing when new species (taxa, community elements) are added to or others proliferate in the community, and decreasing when species drop out or their performance declines.

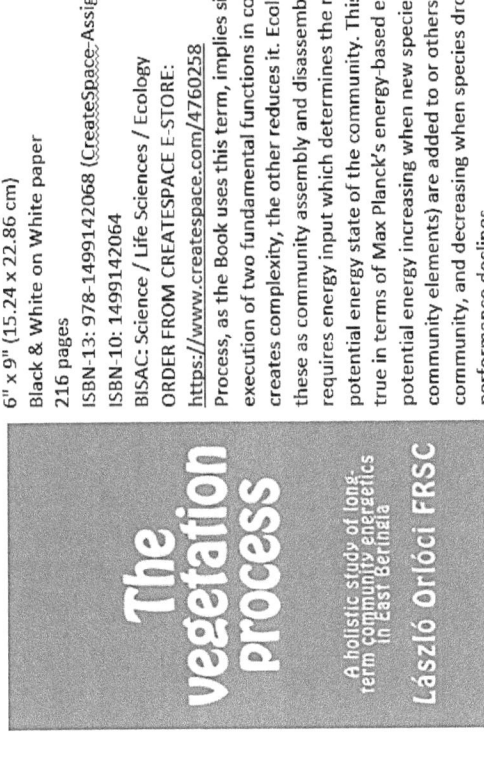

# Quantum analysis of primary succession: The energy structure of a vegetation chronosere in Hawaii Volcanoes National Park

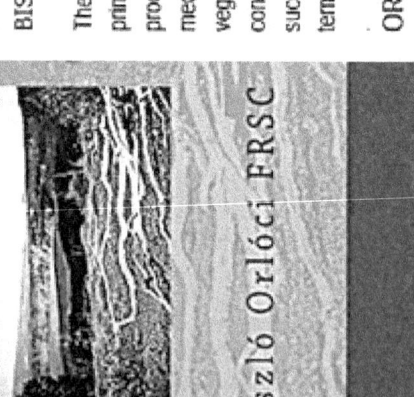

...ored by Laszlo Orlóci FRSC

List Price: **$30.00**

**6" x 9"** (15.24 x 22.86 cm)
Black & White on White paper
54 pages

ISBN-13: 978-1492788997 (CreateSpace-Assigned)
ISBN-10: 1492788996
BISAC: Science / Life Sciences / Ecology

The book revisits the classical idea that the potential energy structure of primary succession is a seamless fusion of foot-prints specific to basic processes which operate on all scales – phylogeny, environmental mediation, and chance. The idea is tested in quantum analysis of a vegetation chronosere in Hawaï Volcanoes National Park. How is the test constructed? What are the outcomes? What do the results tell about primary succession not already known from other sources? Stated in the briefest of terms the test re-quires temporal species performance data...

ORDER FROM CREATESPACE ESTORE:
https://www.createspace.com/4452597

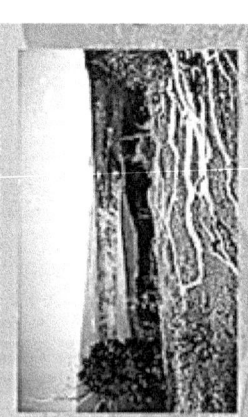

Quantum analysis of primary succession

The energy structure of a vegetation chronosere in Hawai'i Volcanoes National Park

László Orlóci FRSC

## Quantum ecology: Energy structure and its analysis

Authored by László Orlóci FRSC

List Price: $30.00

**6" x 9"** (15.24 x 22.86 cm)
Black & White on White paper
72 pages

ISBN-13: 978-1492183297
ISBN-10: 1492183296
BISAC: Science / Life Sciences / Ecology

Ecology joins forces with quantum theory on the pages of "Quantum Ecology" to create a holistic approach in energy studies.

The infusion of quantum theoretical principles allows the study focus of ecological energetics to shift from the conventional calorific (trophic) flow in ecosystems to the potential energy structure of the vegetation. The books contents cover the theory and techniques in a unique account centred on the energy equation. The equation's component terms define energy footprints specific to ecology's basic processes, such as historic phylogeny, current environmental mediation of transience, and chance. What gives practical value to quantum analysis is its ability to be parameterised by the usual type of survey or experimental data.

The book is offered for classroom use in advanced courses and technical support in research projects.

ORDER FROM CREATESPACE ESTORE:
https://www.createspace.com/4406077

LÁSZLÓ ORLÓCI FRSC

Quantum ecology

Energy structure and its analysis

## Statistical ecology

Like 0

### The quantitative exploration of Nature to reveal the unexpected

**Authored by Laszlo Orlóci Ph.D.**

The book's topics traverse many problem areas in univariate and multivariate data analysis, focussed on current ecological practice. The manner of presentation emphasizes reasoned methodological choices and encourages innovations consistent with the objectives, but mindful of the need to see clearly the regularity conditions which set limits for valid application of statistics in Ecology. The main text is accompanied by external appendices including a technical manual, 47 specialized application programs, and many data files taken from the exercises in the main text. For information please contact: lorloci@uwo.ca

**List Price: $49.90**

Add to Cart

| | |
|---|---|
| **Publication Date:** | Aug 10 2010 |
| **ISBN/EAN13:** | 1453760520 / 9781453760529 |
| **Page Count:** | 372 |
| **Binding Type:** | US Trade Paper |
| **Trim Size:** | 6" x 9" |
| **Language:** | English |
| **Color:** | Black and White |
| **Related Categories:** | Science / Life Sciences / Ecology |

**About the author:**
Orlóci is an INTECOL Distinguished Statistical Ecologist. He is external (academician) Member of the Hungarian Academy of Sciences, and regular (academician) Fellow of the Academy of Sciences of the Royal Society of Canada. He published over 100 papers in scientific journals, numerous monographs and books. His current essays on trajectory analysis, the rules of process governance, and the phylogenetic signal in vegetation transitions have considerable significance for evolutionary ecology and global change science. His present work on energy structures in metacommunities is seminal, pointing to a new direction.

# Statistical multiscaling in dynamic ecology

 Like    0

## Probing the long-term vegetation process for patterns of parameter oscillation

**Authored by László Orlóci Ph.D.**

The Book's conceptualisation of multiscaling theory presents the Next Step in the study of the long-term vegetation process. The context is statistical and the process generating events have proxy in the compositional transitions of the palynological spectra. Familiarity with multiscaling is not a prerequisite. The reader shall learn from the examples how multiscaling techniques helped to identify the self-similar (fractal) nature of the process, isolate low and high instability phases, locate hotspots of compositional transitions, and link these to delayed climatic effects. He or she shall also learn how to gauge process homeomorphy among sites, isolate the random and directed effects found braided into the process, and do much more within a broad yet formal probabilistic framework. The Book's contents are taken in part from a graduate course offered in the Ecology program at UFRGS in Porto Alegre, Brazil. The examples use palynological spectra from sites on the Hungarian Great Plain and in the adjacent Carpathian Mountains. Application programs are available from the author.

List Price: $30.00

**Add to Cart**

**Publication Date:** Mar 15 2012
**ISBN/EAN13:** 1475071388 / 9781475071382
**Page Count:** 96
**Binding Type:** US Trade Paper
**Trim Size:** 6" x 9"
**Language:** English
**Color:** Black and White
**Related Categories:** Science / Life Sciences / Ecology

# Self-organization and mediated transience in plant communities

Like | 0

## What are the rules?

**Authored by Dr. László Orlóci FRSC**

A novel, signal theoretical solution is sketched out for the ecological problem of how to identify and quantitatively express the assembly rules of plant communities. A case study for testing the solution leads to the astonishing conclusion that the phylogenetic signal outperforms the current environmental signal in intensity close to 7 to 1. This indicates high stability and low inclination to environment mediated transience in the community.

**About the author:**

László Orlóci was born in Hungary in 1932. He holds degrees in forest engineering (DFE Sopron), forestry science and biology (BSF, MSc, PhD University of British Columbia), and DSc h.c. in biology (University of Trieste). Orlóci held appointments as NATO Science Fellow (University College of North Wales), professor (University of Western Ontario), and visiting professor at universities in the Americas, the Pacific, Asia, and Europe. He is an INTECOL Distinguished Statistical Ecologists, external (academician) member of the Hungarian Academy of Sciences, and regular Fellow of the Academy of Sciences of the Royal Society of Canada.

**Publication Date:** Nov 11 2011
**ISBN/EAN13:** 1461028221 / 9781461028222
**Page Count:** 70
**Binding Type:** US Trade Paper
**Trim Size:** 6" x 9"
**Language:** English
**Color:** Black and White
**Related Categories:** Science / Life Sciences / Ecology

List Price: $25.00

Add to Cart

# On the energy structure of natural vegetation

In search for community governance rules

**Authored by Laszlo Orloci FRSC**

Briefly about the book …

Vegetation Science meets quantum theory in the energy-based entropy model of this book. The model is based on Max Planck's postulate that potential energy and entropy are interchangeable parameters in resonator complexes. What is a typical outcome of the model in vegetation studies? The model hands users a set of entropy estimates and probabilities based on which the strength and uniqueness of a metacommunity's energy structure can be characterised in comparative terms.

**About the author:**

Orlóci is an INTECOL Distinguished Statistical Ecologist. He is external (academician) Member of the Hungarian Academy of Sciences, and regular (academician) Fellow of the Academy of Sciences of the Royal Society of Canada. Orlóci published over 100 papers in scientific journals, numerous monographs, books and book chapters. His current essays on trajectory analysis, the rules of process governance, and the phylogenetic signal in vegetation transitions have considerable significance for evolutionary ecology and global change science. His present work on energy structures in metacommunities is seminal, pointing to a new direction.

List Price: $30.00

**Add to Cart**

| | |
|---|---|
| **Publication Date:** | Jan 30 2013 |
| **ISBN/EAN13:** | 1482319373 / 9781482319378 |
| **Page Count:** | 46 |
| **Binding Type:** | US Trade Paper |
| **Trim Size:** | 6" x 9" |
| **Language:** | English |
| **Color:** | Black and White |
| **Related Categories:** | Science / Life Sciences / Ecology |

# Flexible computing in statistical ecology  [f] Like  0

## External appendix to accompany L. Orlóci's Statistical Ecology

**Authored by Dr. László Orlóci**

Problem flexible computing in statistical ecology

The Book describes more than 40 executable (.exe) computer programs and presents examples of application which correspond to the examples included in Statistical Ecology*. The programs are flexibly problem specific and conversational. They allow option-driven selective access to specific statistical tasks. Linkages are possible through standard output and input. The description includes in each case a brief introduction, a record of the start up dialogue, and detailed record input and output sets. The source code is in True Basic. The programs are compiled and linked on a 32 bit Windows XP system and tested up to Windows 7.
The executable program library, data files and a note to users are distributed free of charge to purchasers of the Technical Manual. Requests for download information should be directed to the URL address lorloci@uwo.ca. Prior to running the application programs, installation of a recent version of True Basic (see Internet for sources) on the user's system is strongly advised.
* Orlóci, L. 2010. Statistical Ecology. The quantitative exploration of nature to reveal the unexpected. Scada Publishing, Online Edition. Copies are available from the distributor
https://www.createspace.com/3476529

| | |
|---|---|
| **Publication Date:** | Apr 05 2011 |
| **ISBN/EAN13:** | 1460972953 / 9781460972953 |
| **Page Count:** | 142 |
| **Binding Type:** | US Trade Paper |
| **Trim Size:** | 6" x 9" |
| **Language:** | English |
| **Color:** | Black and White |
| **Related Categories:** | Science / Life Sciences / Ecology |

List Price: $30.00

**Add to Cart**

Statistical Ecology, A reasoned approach.

# Reader's notes

# Biographic notes …

László Orlóci was born into a military family in Hungary in 1932. He holds degrees in forest engineering (DFE Sopron), forestry science and biology (BSF, MSc, PhD University of British Columbia), and DSc *h.c.* in biology (University of Trieste); held appointments as NATO Science Fellow (University College of North Wales), professor (University of Western Ontario), and visiting professor at universities in the Americas, the Pacific, Asia, and Europe. He is INTECOL's Distinguished Statistical Ecologists, external academician member of the Hungarian Academy of Sciences, and regular Fellow of the Academy of Sciences of the Royal Society of Canada.

Orlóci published over 100 papers in scientific journals, numerous monographs and books. His current monographs address multiscale trajectory analysis, the rules of process governance, the phylogenetic signal in vegetation transitions, and fundamental problems in the theory and *modus operandi* of quantum ecology.

Orlóci is married to Márta Mihály, a Sopron forest engineering alumna. Their daughter Martha is a Geography graduate of Western University in Canada. He has two granddaughter. Kathryn Orlóci-Goodison is third year environmental resources management major and Ruth Orlóci-Goodison is first year Presidential Scholar in biology, both at Lakehead University in Thunder Bay, Ontario.

2015-07-28

www.ingramcontent.com/pod-product-compliance
Lightning Source LLC
Chambersburg PA
CBHW071015180526
45168CB00003B/1427